ANIMALS OF THE TUNDRA

MIGHTY MUSK OXEN

THERESA EMMINIZER

PowerKiDS press

New York

Published in 2022 by The Rosen Publishing Group, Inc.
29 East 21st Street, New York, NY 10010

Copyright © 2022 by The Rosen Publishing Group, Inc.

All rights reserved. No part of this book may be reproduced in any form without permission in writing from the publisher, except by a reviewer.

First Edition

Portions of this work were originally authored by Roman Patrick and published as *Musk Oxen*. All new material in this edition authored by Theresa Emminizer.

Editor: Jill Keppeler
Book Design: Michael Flynn

Photo Credits: Cover, p. 1 Bjørn H Stuedal/500px/Getty Images; (series background) Stefan Sorean/Shutterstock.com; p. 5 sarkophoto/iStock/Getty Images; p. 7 Anagramm/iStock/Getty Images; p. 9 (bull) AlbyDeTweede/iStock/Getty Images; p. 9 (cow) Arterra/Universal Images Group/Getty Images; p. 11 Robert Haasmann/Getty Images; p. 13 Frank Fichtmüller/iStock/Getty Images; p. 15 Franz Aberham/Photographer's Choice RF/Getty Images; p. 17 Betty Shelton/Shutterstock.com; p. 19 Fred Bruemmer/Photolibrary/Getty Images Plus; p. 21 Giedriius/Shutterstock.com.

Library of Congress Cataloging-in-Publication Data

Names: Emminizer, Theresa, author.
Title: Mighty musk oxen / Theresa Emminizer.
Description: New York : PowerKids Press, [2022] | Series: Animals of the
   tundra | Includes index.
Identifiers: LCCN 2020021703 | ISBN 9781725326392 (library binding) | ISBN
   9781725326378 (paperback) | ISBN 9781725326385 (6 pack)
Subjects: LCSH: Muskox–Juvenile literature. | Tundra animals–Juvenile
   literature.
Classification: LCC QL737.U53 E483 2022 | DDC 599.64/78–dc23
LC record available at https://lccn.loc.gov/2020021703

Manufactured in the United States of America

Some of the images in this book illustrate individuals who are models. The depictions do not imply actual situations or events.

CPSIA Compliance Information: Batch #CSPK22. For Further Information contact Rosen Publishing, New York, New York at 1-800-237-9932.

CONTENTS

Big and Stinky! 4
Huge and Hairy 6
Big Horns 8
A Warm Winter Coat 10
Moving in Herds 14
Finding Food 16
Standing Together 18
Musk Oxen in Danger 20
Fancy Features 21
Glossary 22
For More Information 23
Index 24

Big and Stinky!

Musk oxen are big, hairy animals that live in the **tundra**. They're also stinky! They give off a very strong smell because of a matter they produce. This matter is called musk. That's how they got their name.

Huge and Hairy

Musk oxen hair can be as long as 35 inches (88.9 cm) on most of their body! Sometimes it nearly sweeps the ground. **Male** musk oxen, or bulls, can weigh up to 900 pounds (408.2 kg).

Big Horns

Musk oxen have a big head with two sharp horns. Bulls have longer horns. **Female** oxen, or cows, have smaller horns. Musk oxen use their horns as **weapons** against **predators**.

A Warm Winter Coat

A musk ox always has a long coat of hair. In the winter, it has two coats! There's a thick, softer coat of **wool** under the long, brown coat. These coats help keep musk oxen warm in tundra winters.

In spring, musk oxen don't need two coats to stay warm. They start to lose the inner coat. They look kind of funny! They'll regrow their winter coat when it starts to get cold again.

Moving in Herds

Musk oxen live in big groups called herds. Each herd has about 20 to 30 oxen. They're led by one female ox. They move over the tundra, looking for food. They eat plants, including seeds and berries.

Finding Food

In the winter, musk oxen use their hooves to uncover roots and mosses to eat. In the summer, they also eat grasses and flowers. They build up fat to help them in winter.

Standing Together

Musk oxen have predators, including wolves and bears. If there's danger, adult musk oxen form a circle around the young ones. They'll use their horns to fight. Smaller herds might form a line with the young behind it.

Musk Oxen in Danger

The musk ox's most common predators are humans. Hunters have killed many musk oxen for their hides and meat. The animals nearly died out. Today, laws keep musk oxen safe in some parts of the world.

Fancy Features

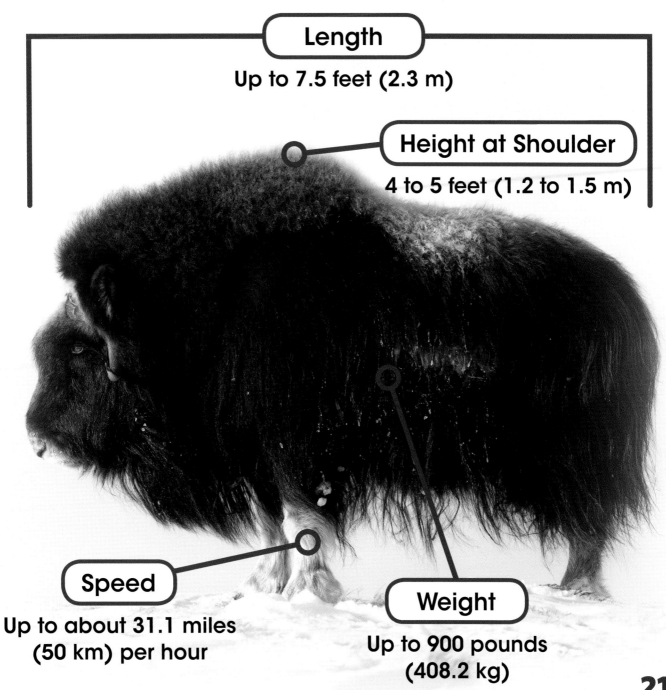

Length
Up to 7.5 feet (2.3 m)

Height at Shoulder
4 to 5 feet (1.2 to 1.5 m)

Speed
Up to about 31.1 miles (50 km) per hour

Weight
Up to 900 pounds (408.2 kg)

GLOSSARY

female: A girl.

male: A boy.

predator: An animal that hunts other animals for food.

tundra: Cold northern lands that lack forests and have permanently frozen soil below the surface.

weapon: A tool used to fight someone or something.

wool: Soft, thick hair on some animals.

FOR MORE INFORMATION

WEBSITES

Musk Ox
www.nationalgeographic.com/animals/mammals/m/musk-ox/
Learn more about the adaptations and behavior of the mighty musk ox on this National Geographic website.

Ovibos moschatus
animaldiversity.org/accounts/Ovibos_moschatus/
The University of Michigan Museum of Zoology presents a wealth of information on the musk ox on this in-depth website.

BOOKS

Alkire, Jessie. *Musk Oxen.* Minneapolis, MN: Abdo Publishing, 2019.

Cocca, Lisa Colozza. *Tundra Animals.* North Mankato, MN: Rourke Educational Media, 2019.

Publisher's note to parents and teachers: Our editors have reviewed the websites listed here to make sure they're suitable for students. However, websites may change frequently. Please note that students should always be supervised when they access the internet.

INDEX

B
bears, 18
bulls, 6, 8, 9

C
coat, 10, 12
cows, 8, 9

F
food, 14, 16

H
hair, 6, 10
head, 8
herds, 14, 18
hooves, 16
horns, 8, 18

M
musk, 4

P
predators, 8, 18, 20

S
smell, 4
spring, 12
summer, 16

W
winter, 10, 12, 16
wolves, 18
wool, 10